THE FLOOD

THE FLOOD

CHARLES TOMLINSON

Oxford New York Toronto Melbourne
OXFORD UNIVERSITY PRESS
1981

Oxford University Press, Walton Street, Oxford OX2 6DP

London Glasgow New York Toronto
Delhi Bombay Calcutta Madras Karachi
Kuala Lumpur Singapore Hong Kong Tokyo
Nairobi Dar es Salaam Cape Town
Melbourne Wellington
and associate companies in
Beirut Berlin Ibadan Mexico City

British Library Cataloguing in Publication Data
Tomlinson, Charles
The flood.
I. Title
821'.914 PR6039.0349F/ 80-41887
ISBN 0-19-211944-3

Set by King's English Typesetters Ltd., Cambridge
Printed in Great Britain
at the University Press, Oxford
by Eric Buckley
Printer to the University

To Brenda

Acknowledgements

Acknowledgements are due to the editors of the following periodicals and books in which some of these poems first appeared: *Agenda, The Christian Science Monitor, Escandalar, The Hudson Review, The London Magazine, The London Review of Books, The New Statesman, Observer Magazine, The Paris Review, Poetry Nation Review, South West Review, Stand, The State of the Language* (ed. Michaels and Ricks, University of California Press), *Vanderbilt Poetry Review*.

I must also thank the producer of 'Poetry Now' (BBC). 'Poem' was first published by Matrix Press, Palo Alto, as Chimera Broadside Two.

Contents

Snow Signs

They say it is waiting for more, the snow
   Shrunk up to the shadow-line of walls
In an arctic smouldering, an unclean salt,
   And will not go until the frost returns
Sharpening the stars, and the fresh snow falls
   Piling its drifts in scallops, furls. I say
Snow has left its own white geometry
   To measure out for the eye the way
The land may lie where a too cursory reading
   Discovers only dip and incline leading
To incline, dip, and misses the fortuitous
   Full variety a hillside spreads for us:
It is written here in sign and exclamation,
   Touched-in contour and chalk-followed fold,
Lines and circles finding their completion
   In figures less certain, figures that yet take hold
On features that would stay hidden but for them:
   Walking, we waken these at every turn,
Waken ourselves, so that our walking seems
   To rouse some massive sleeper out of winter dreams
Whose stretching startles the whole land into life,
   As if it were us the cold, keen signs were seeking
To pleasure and remeasure, repossess
   With a sense in the gathered coldness of heat and height.
Well, if it's for more the snow is waiting
   To claim back into disguisal overnight,
As though it were promising a protection
   From all it has transfigured, scored and bared,
Now we shall know the force of what resurrection
   Outwaits the simplification of the snow.

Their Voices Rang

Their voices rang
through the winter trees:
they were speaking and yet it seemed they sang,
the trunks a hall of victory.

And what is that and where?
Though we come to it rarely,
the sense of all that we might be
conjures the place from air.

Is it the mind, then?
It is the mind received,
assumed into a season
forestial in the absence of all leaves.

Their voices rang
through the winter trees and time
catching the cadence of that song
forgot itself in them.

The Double Rainbow

To Ulalume Gonzalez de Leon

When I opened your book
a rainbow shaft
looked into it
through the winter window:

a January light
searching the pane
paused there refracted
from white onto white:

so words become
brides of the weather
of the day in the room
and the day outside:

in the light of the mind
the meanings loom
to dance in their own
glimmering spectrum

For Miriam

I

I climbed to your high village through the snow,
   Stepping and slipping over lost terrain:
Wind having stripped a dead field of its white
   Had piled the height beyond: I saw no way
But hung there wrapped in breath, my body beating:
   Edging the drift, trying it for depth,
Touch taught the body how to go
   Through straitest places. Nothing too steep
Or narrow now, once mind and muscle
   Learned to dance their balancings, combined
Against the misdirections of the snow.
   And soon the ground I gained delivered me
Before your smokeless house, and still
   I failed to read that sign. Through cutting air
Two hawks patrolled the reaches of the day,
   Black silhouettes against the sheen
That blinded me. How should I know
   The cold which tempered that blue steel
Claimed you already, for you were old.

II

Mindful of your death, I hear the leap
   At life in the *resurrexit* of Bruckner's mass:
For, there, your hope towers whole:
   Within a body one cannot see, it climbs
That spaceless space, the ear's
   Chief mystery and mind's, that probes to know
What sense might feel, could it outgo
   Its own destruction, spiralling tireless
Like these sounds. To walk would be enough
   And top that rise behind your house
Where the land lies sheer to Wales,
   And Severn's crescent empties and refills
Flashing its sign inland, its pulse
   Of light that shimmers off the Atlantic:

4

For too long, age had kept you from that sight
　And now it beats within my eye, its pressure
A reply to the vein's own music
　Here, where with flight-lines interlinking
That sink only to twine and hover the higher,
　A circling of hawks recalls to us our chains
And snow remaining hardens above your grave.

III

You wanted a witness that the body
　Time now taught you to distrust
Had once been good. 'My face,' you said –
　And the Shulamite stirred in decembering flesh
As embers fitfully relit – 'My face
　Was never beautiful, but my hair
(It reached then to my knees) was beautiful.'
　We met for conversation, not conversion,
For you were that creature Johnson bridled at –
　A woman preacher. With age, your heresies
Had so multiplied that even I
　A pagan, pleaded for poetry forgone:
You thought the telling-over of God's names
　Three-fold banality, for what you sought
Was single, not (and the flame was in your cheek)
　'A nursery rhyme, a jingle for theologians.'
And the incarnation? That, too, required
　All of the rhetoric that I could bring
To its defence. The frozen ground
　Opened to receive you a slot in snow,
Re-froze, and months unborn now wait
　To take you into the earthdark disincarnate.

IV

A false spring. By noon the frost
　Whitens the shadows only and the stones
Where they lie away from light. The fields
　Give back an odour out of earth

Smoking up through the haysmells where the hay
    – I thought it was sunlight in its scattered brightness –
Brings last year's sun to cattle wintering:
    The dark will powder them with white, and day
Discover the steaming herd, as beam
    On beam, and bird by bird, it thaws
Towards another noon. *Et resurrexit:*
    All will resurrect once more,
But whether you will rise again – unless
    To enter the earthflesh and its fullness
Is to rise in the unending metamorphosis
    Through soil and stem . . . This valediction is a requiem.
What was the promise to Abraham and his seed?
    That they should feed an everlasting life
In earthdark and in sunlight on the leaf
    Beyond the need of hope or help. But we
Would hunger in hope at the shimmer of a straw,
    Although it burned, a mere memory of fire,
Although the beauty of earth were all there were.

V

In summer's heat, under a great tree
    I hear the hawks cry down.
The beauty of earth, the memory of your fire
    Tell of a year gone by and more
Bringing the leaves to light: they spread
    Between these words and the birds that hang
Unseen in predatory flight. Again,
    Your high house is in living hands
And what we were saying there is what was said.
    My body measures the ground beneath me
Warm in this beech-foot shade, my verse
    Pacing out the path I shall not follow
To where you spoke once with a wounded
    And wondering contempt against your flock,
Your mind crowded with eagerness and anger.
    The hawks come circling unappeasably. Their clangor
Seems like the energy of loss. It is hunger.
    It pierces and pieces together, a single note,

The territories they come floating over now:
  The escarpment, the foreshore and the sea;
The year that has been, the year to be;
  Leaf on leaf, a century's increment
That has quickened and weathered, withered on the tree
  Down into this brown circle where the shadows thicken.

The Recompense

                    The night of the comet,
         Sunset gone, and shadow drawing down
Into itself landscape, horizon, sky,
         We climbed the darkness. Touch
Was all we had to see by, as we felt
         For a path among the crowding trees:
Somehow, we threaded them, came through
         At last to the vantage we had aimed for:
It was viewless: a sole star,
         The cold space round which seemed
The arena a comet might be found
         Sparkling and speeding through, if only
One waited long enough. We waited.
         No comet came, and no flame thawed
The freezing reaches of our glance: loneliness
         Quelled all we saw – the wide
Empire of that nightworld held
         To the sway of centuries, sidereal law,
And the silent darkness hiding every star
          Save one: had we misheard the date?
A comet, predicted, might be late
         By days perhaps? Chilled, but unwillingly
We took the tree-way down; and ran
         Once feet, freed from obstruction,
Could feel out the smoothest path for us
         From wood to warmth. Now that we faced away
From the spaces we had scanned for light,
         A growing glow rose up to us,
Brought the horizon back once more
         Night had expunged: it travelled contrary
To any comet, this climbing brightness:
         We wound the sight towards us as we went,
The immense circle of the risen moon
         Travelling to meet us: trees
Wrote themselves out on sea and continent,
         A cursive script where every loop and knot
Glimmered in hieroglyph, clear black:
         We – recompense for a comet lost –

Could read ourselves into those lines
　　Pulsating on the eye and to the veins,
Thrust and countercharge to our own racing down,
　　Lunar flights of the rooted horizon.

Poem

It falls onto my page like rain
the morning here
and the ink-marks run
to a smoke and stain, a vine-cord, hair:

this script that untangles itself
out of wind, briars, stars unseen,
keeps telling me what I mean
is theirs, not mine:

I try to become all ear
to contain their story:
it goes on arriving from everywhere:
it overflows me

and then:
a bird's veering
into sudden sun
finds me for a pen

a feather on grass,
a blade tempered newly
and oiled to a gloss
dewless among dew:

save for a single
quicksilver drop –
one from a constellation
pearling its tip

In April

I thought that the north
wind was treating the wood
as a thunder-sheet it was
thunder itself had
merged with the roar
of the air in a vast
voice a judgement chord
and the winter that would not go
was blocking spring
through the upper sky piling
ledges of cold onto
ridges of ripening warmth
quaking across the entire
expanse and pushing sun
back into a livid
pre-world light as it
rolled end and beginning
up in a single emphatic
space-travelling verb breath word

The Order of Saying

'As soon as the blackthorn comes in flower
    The wind blows cold,' she says:
I see those bushes tossed and whitening,
    Drawing the light and currents of the air
Into their mass and depth; can only see
    The order of her saying in that flare
That rises like a beacon for the wind
    To flow into, to twist and wear
Garment and incandescence, flag of spring.

The Lesson

The larks, this year,
fly so early and so high,
it means, you tell me, summer
will be dry and hot,
and who am I
to gainsay that prophecy?
For twenty years here have not
taught me to read with accuracy
the signs either of earth or sky:
I still keep the eye of a newcomer,
a townsman's eye:
but there is time yet
to better my instruction
in season and in song:
summer on summer

Hay

The air at evening thickens with a scent
That walls exude and dreams turn lavish on –
Dark incense of a solar sacrament
Where, laid in swathes, the field-silk dulls and dries
To contour out the land's declivities
With parallels of grass, sweet avenues:
Scent hangs perpetual above the changes,
As when the hay is turned and we must lose
This clarity of sweeps and terraces
Until the bales space out the slopes again
Like scattered megaliths. Each year the men
Pile them up close before they build the stack,
Leaving against the sky, as night comes on,
A henge of hay-bales to confuse the track
Of time, and out of which the smoking dews
Draw odours solid as the huge deception.

The Conspiracy

My writing hand moves washed in the same light now
   As the beasts in the field beneath this window:
The hot day hangs in the glitter and the shade:
   Those toys of Arden, seeing and half-seeing,
The rocking of the leaves, the slow berceuse,
   The airs advancing to invade the heat
But warming to a world of ease, all
   Breathe together this conspiracy:
Butterflies tongue the nectaries and summer
   Is hiving time against the centuries.

The Gate

Someone has set up a gate here
  In this unfenced field. Waiting for its fence
It teases the sight. Is it that one feels half-blind,
  The mind demanding an enclosure that the eye
Cannot supply it with? Or is it an X-ray eye
  One has, melting wall to a nothing,
And the grass greening up at one, that returns
  A look with intimations of a place unspaced
And thus not quite there? The mocked mind,
  Busy with surroundings it can neither bound nor unbind,
Cedes to the eye the pleasure of passing
  Where, between the gate's five bars,
Perpetual seawaves play of innumerable grasses.

At the Edge

The offscape, the in-folds, secreted
   Water-holes in the boles of trees,
Abandoned bits, this door of water
   On the wood's floor (knock with the breath
And enter a world reverted, a catacomb
   Of branching ways where the roots splay):
Edges are centres: once you have found
   Their lines of force, the least of gossamers
Leads and frees you, nets you a universe
   Whose iridescent weave shines true
Because you see it, but whose centre is not you:
   Through the wheel of a web today I saw
The wren, that mere mouse of a bird
   Hurry from its hole and back again
With such an energy of glancing lightness
   It made me measure all the force unspied
That stirred inside that bank, still
   As it seemed, beside the flashing watercourse
That came straight on contrary to my direction and
   Out of the dereliction of an edge of woodland.

The Near and the Far

It is autumn and there are no flowers in this desert garden. Open to the breeze and to the light, it gauges near against far and resolves them. For the breeze, that pushes to and fro the shadows of the hanging pods of the catalpa tree, is stirring the dust over hundreds of desert miles, lifting it in a veil before the long mass of the Sandía Mountains and picking to pieces the vapour trails of the jets that scar the high and intense blue. As for the light: it reveals sandy distances, ripening the tints, as shadows lengthen, to dusty reds and deep yellows. Here, in the eyes' immediate reaches, it brings out each facet of bark still armouring the stakes that fence the dry yard. Yet they do not fence it in: the endless streaming of light through the cottonwoods, whose leaves are sapless but dazzling, reaches us from an unseen source, a blinding immensity. The sky is full of crossroads of jet-trails, downy tracks gigantically decaying. At foot level, ants are busy in their moonworld of dust and craters.

This desert garden lies at the town's edge, open to the breath of a thousand miles. The engine of a passing truck grinds by with the silence of desert spaces first surrounding then absorbing it. A train passes. Its siren seems to sound out the distances it is entering and, as it were, creating. Waves of wind are rolling across the sky. And from the cottonwoods come the rattle of their leaves, crepitations, lulls, whispers, splashings, like a sea breaking against shore and jetty, where no shore or jetty ever were, on this dried-up bed of a vanished ocean.

Rooted in dust, what do they feed on, these great trees, whose abundance of falling leaves attests the fullness of their summer? In this desert garden there is no succulence. The spiny plants catch at one's ankles, one's shoes are perpetually powdered over. Cottonwood leaves leap across the ground in the wind, side by side with the cicadas that land with a metallic crashing sound among the already shed foliage.

Red ants, dwarfed by their own gangling shadows, carry high their polished abdomens, rushing out on some frantic mission through the warm dust and its maze of cat- and dogpaw prints, shoeshapes, trails where the hose has been dragged across it. The ant nests are composed of a circular wall of dust like a tiny Celtic fortification raised on the desert. Cruzita will not kill the ants because they are part of a divine and unsoundable dispensation.

However, she has stood empty jam jars up to their necks in sand, so that the ants can fall into them if that dispensation permits them to. She, the tutelary spirit of this house and garden, has brought into their enclosure the memory of that wisdom which has made life possible across desert and mesa.

She has lit her morning fire. The piñon of which it is built sends out an upland fragrance of pinewoods over the ragged edges of the suburb. The smoke tells the nostrils of the rarer, cooler autumnal air in those fastnesses, recalls to the mind's eye the mesa villages and the sight of trees growing along the mesa top that, seen from afar, look like the stubble of a beard. The light-blue, rising smoke winding above the house, spreading, perfuming the atmosphere, brings to a lingering resolution in its chord of colour, the blue shadows exploring the heights, the dust-haze that the breeze has lifted and a sense of possibility that, investing space, pours into this garden with the light and air.

Albuquerque

Driving once more
on Central – though what
it is central to
now the town has spread

up to the foothills –
I caught one thing
out of the past that
still was there –

the Kimo Kinema
built in my birth year –
tile, fresco and cement-adobe –
style, tribal art-deco:

Corbusier
could scarcely have applauded
architecture so little tectonic
or so tawdry

and yet a
pupil of his
aware that this
is neither Ronchamps nor Sabbioneta

is actually here
restoring that memory
of a gone year:
I am antique already

Jemez

When we were children said
Eva they told us
the trees on the skyline there
– we turned to see
the trees on the skyline there
stand staring down –
were the kachinas and we
believed them but today
if you say to children
the trees on the skyline there
– the skyline trees stand
calling up sap out of rock and sand –
are the kachinas they
reply kachinas?
they're nothing but trees

kachinas: tribal spirits

Cochiti

The cries
of the eagle
dancers at Cochiti
rise to such a
complete complicity
with the way it is
when eagles
speak what they have to say
that the sky
brings down
over thinning snow
unfurling the black
line of wings
slackening then
tautening them back
to take the thrust
of the wind and mount it
to wander away
above the pocked snow-whitenesses
of plain hills mesas
an eagle

Abiquiu

Rattlesnakes I've had them
in every room of this house
she said: one
lay there suspended
– it might have been in mid-air –
on the down of a piled rug
as if it were swimming:
then one night
I opened the patio
door and the cat
sat there faced
by a snake: neither
moved but the snake kept up
its warning rattle:
day or night
I wouldn't step
right out there
onto the patio
without looking first:
do you know
it's not from that
tight coil
that a rattler strikes:
it's – she twisted her fingers together –
from a tangle
that has no shape
but I tell you
whatever its reputation
a rattlesnake is a gentleman:
before he strikes he
always lets you know

Quarai

Two dogs
wait for the crumbs
from our desert meal:
they are not importunate
but thoughtful in their anticipation:
I recognize their pondering courtesy
in the custodian – they belong to him –
opening his show for us
– artefacts and a plan
of the vanished pueblo –
under a leaking roof
which come spring days
he must he says mend:
La Purissima Concepcion
where the inquisition spied
for witches and blasphemers
cannot account for him:
it died and dwindled
into sky-roofed sandstone here
through pestilence, famine
drought and massacre
unprepared for this
accommodating gentleness
that says: You
hurry on back
whenever you want to

Under the Bridge

Where the ranch-house disappeared its garden
seeded and the narcissi
began through a slow mutation
to breed smaller and smaller stars
unimpaired in scent: beside these
the horns of the cala lilies
each scroll protruding an insistent
yellow pistil seem from their scale
and succulent whiteness to belong
to an earlier world:
if there were men in it the trellises
that brace these stanchions
would fit the scale
of their husbandry and
if they made music it would
shudder and rebound
like that which travels down
the metal to the base
of this giant instrument
bedded among teazle, fennel, grass
in a returning wilderness
under the bridge

San Francisco

Cronkhite Beach

They look surreal
you said: you meant
the figures along the cliff
above us that the evening light
was thrusting forward in silhouette as if
the sheer stance of curiosity
– they were all stillness facing out to sea –
had magnified them: wave
after wave was entering
from the horizon: moving mountains
which as they sailed-in threw
the surf backwards from their peaks
each like a separate volcano:
and the diminishing force unspent
you could measure by the pitch
of the off-shore buoy
belling the shallows as the sea
tugged past it to become
white on the darkening beach
a frayed rope of foam: and there
were the peaceful Brobdignagians still
stationed out massively along that hill:
tall as the objects of their contemplation
they looked surreal: so real

Poolville, N.Y.

Brekekekex?
The frogs of upstate speak
with a mellifluousness as ungreek
as their names – 'spring peepers':
neither hoarse nor flatulent
like the ones elsewhere, this
breed were never peasantry
like those in Ovid's metamorphosis,
but nature's aristocracy –
'We are all princes here':
and thus will never ask
for a log for king
and end up with a stork:
they must sing four nights and then
they've safely brought the springtime in
past tardy drifts and up the trees
to a music that outdates
Ovid and Aristophanes.

Parsnips

for Ted Chamberlin

A mixed crop. I dig a clump:
Crotches seamed with soil,
Soil clinging to every hair,
To excrescences and mandrake mandibles.
Poor bare forked earth-stained animals,
One comes up whole and white,
A vegetable Adam. I take the lot
And wash them at the stream.
Rubbing, rinsing, I let fall
Inevitably this image of perfection,
Then rush for a garden rake
To fish it back again and run
Trying to out-race the current's
Rain-fed effervescence:
Fit image of the poet, he
In the waterproof, with the iron comb who goes
Hunting a prey that's halfway to the sea.

Programme Note

Reading this, you are waiting for the curtain
    To go up on a glade, vistaed valley
Or colonnade of lath. Yet you are not here
    To view a painting – the painted thing
Like the written word, is there for the hearing –
    To which end the tympanist stretches his ear
To interrogate a drumskin, hangs over
    Undistracted by bell note or forest murmur
In horn and harp. Cellist pursues
    An intent colloquy with his instrument,
Urging nerve and string up to that perfection
    He may falter at. For, the aria done,
It is he alone who must comment on
    The meaning of it, and bars he is testing now
Climb then on a faultless bow
    Out of the darkened pit as the hero pauses
To resume in song. He, too, unseen
    Sweating into his paint, runs through
(In mind, that is) the perils of a part
    That from start to finish (and this is true
Of every bubble and iota of these tuning notes)
    Raises its fragment to build a single arc
Of sound. Yet suppose that you are here
    Tonight to share in the good conscience
Of all masquerade, that this wood or square
    Waits to be filled with cadences in which
Taking leave of the humid north,
    The steam of Niebelheim, you find yourself
With time and light enough to feel
    The filigree of things, and dare to be
Superficial out of profundity: suppose
    This composer of yours had for a beginning
The merest ravelled thread of a plot –
    The sort of thing a poet would wince at
Or a bad poet write –, it flows
    Out from his musician's mind, not

As the Gesammtkunstwerk (let that dragon sleep),
    The streambed's deep self-inspection,
But the purest water where reflection
    Pooling for a moment, is drawn along
Over drops and through recesses, to emerge
    Strong though contained, a river of song:
You feel that you could leap it from side to side:
    Its dazzle and deftness so take hold,
They convince the mind that it might be
    Equally agile, equally free:
Are you the swift that dips here, or the course
    Of sheer, unimpeded water, the counterforce
Of rock and stone? But images lie –
    Not the Ding-an-sich, but the light to see it by:
And no river could convey the artifice
    And no landscape either, the pulse of this:
A closer thing, it is as if thought might sing
    To the bloodbeat, set it racing;
As if . . . and yet a man shaped this
    Who read the fragile story from the start
As that which his art would make of it –
    So that, in the mind, the body dances
To this flowing fiction – soprano, tenor
    And basso buffo – believing all,
Limpid, unpsychological: and finding true
    A wholly imaginary passion– passion spaced until
Meted and metred-out, its urgency
    Does not merely billow up to fill
The gallery with sound. But I have said
    Enough and the musicians mouthed and bowed
Their accordant A; the light glows out
    On the gildings, and here is the man in black
And white who holds this world of yellow and green and red
    Together, and his first chord cuts the last whisper back.

On a Pig's Head

Once it had gorged itself
to a pitch of succulence, they slew it:
it was the stare in the eyes
the butcher hated, and so removed
with a quick knife,
transforming the thing
to a still life, hacked
and halved, cross-cutting it
into angles with ears.
It bled no more,
though the black pearls
still lurked on its rawness
The ears were streaked with wax,
the teeth stained near the roots
like an inveterate smoker's.
It was the nose looked freshest –
a rubbery, soft pink.
With a spill of paper, I cleaned
the orifice of each ear,
and played water into the nostrils.
The brain was a mere thimble of brain,
and the tongue, smaller than a sheep's
sliced neatly. The severed ears
seemed delicate on their plate
with their maze of veins.
When we submerged it in brine
to change it to brawn and galantine,
it wouldn't fit the bowls:
evidently, it had been conceived
for a more capacious age.
Divided, it remained massive
leaving no room for reflection
save that peppercorns, cloves
of garlic, bay-leaves and wine
would be necessary for its transformation.

When set to boil, it required
a rock, a great
red one
from Macuilxochitl
to keep it down.

Ritornello

Wrong has a twisty look like wrung misprinted
Consider! and you con the stars for meaning
Sublime comes climbing from beneath the threshold
Experience? you win it out of peril
The pirate's cognate. Where did the words arise?
Human they sublimed out of the humus
Surprised by stars into consideration
You are wrung right and put into the peril
Of feelings not yet charted lost for words
Abstraction means something pulled away from
Humus means earth place purchase and return

San Fruttuoso:
the divers

Seasalt has rusted the ironwork trellis
at the one café. Today
the bathers are all sun-bathers
and their bodies, side by side,
hide the minute beach:
the sea is rough and the sun's
rays pierce merely fitfully
an ill-lit sky. Unvisited,
the sellers of lace and postcards
have nothing to do, and the Dorias
in their cool tombs under the cloisters
sleep out history unfleshed.
Oggi pesce spada
says the café sign, but we
shall eat no swordfish today:
we leave by the ferry
from which the divers are arriving.
We wait under an orange tree
that produces flowers but no oranges.
They litter the rocks with their gear
and begin to assume
alternative bodies, slipping
into black rubbery skins with *Caution*
written across them.
They are of both sexes. They strap on
waist weights, frog feet,
cylinders of oxygen,
they lean their heaviness which water will lighten
back against rock, resting there
like burdened seals.
They test their cylinders
and the oxygen hisses at them.
They carry knives
and are well equipped to encounter
whatever it is draws them downwards
in their sleek black flesh.

The postcards show Christ –
Cristo del mare –
sunk and standing on his pedestal
with two divers circling
as airy as under-water birds
in baroque, ecstatic devotion
round the bad statue.
Will they find calm down there
we wonder, stepping heavily
over the ship-side gap,
feeling already the unbalancing
pull of the water under us.
We pass the granular rocks
faulted with long scars.
The sea is bristling up to them.
The straightness of the horizon
as we heave towards it
only disguises the intervening
sea-roll and sea-chop, the clutching glitter.
I rather like
the buck of the boat. What I dislike
with the sea tilting at us
is the thought of losing one's brains
as one slides sideways
to be flung at the bulwarks
as if weightless, the 'as if'
dissolving on impact
into bone and blood.
The maternal hand tightens
on the push-chair
that motion is dragging at:
her strapped-in child is asleep.
Perhaps those invisible divers –
luckier than we are –
all weight gone
levitate now
around the statue,
their corps de ballet
like Correggio's sky-
swimming angels, a swarm
of batrachian legs:

they are buoyed up by adoration,
the water merely an accidental aid
to such staggeringly
slow-motion pirouettes
forgetful of body, of gravity.
The sea-lurch snatches
and spins the wheels of his chair
and the child travels the sudden gradient
caught at by other hands,
reversed in mid-flight
and returned across the up-
hill deck to his mother:
a visitor,
she has the placid
and faintly bovine look
of a Northern madonna
and is scarcely surprised; he, too,
stays perfectly collected
aware now of what it was he had forgotten
while sleeping – the stuff
he was chewing from a packet,
which he continues to do.
He has come back to his body once more.
How well he inhabits his flesh:
lordly in unconcern,
he is as well accoutred as those divers.
He rides out the storm chewing and watching,
trustfully unaware
we could well go down
– though we do not, for already
the town is hanging above
us and the calm quay water.
From the roofs up there
perhaps one could see the divers
emerging, immersing,
whatever it is they are at
as we glide forward
up to the solid, deck to dock,
with salted lips.

That same sea
which wrecked Shelley
goes on rocking behind
and within us, hiding
its Christ, its swordfish,
as the coast reveals
a man-made welcome to us
of wall, street, room,
body's own measure and harbour,
shadow of lintel, portal
asking it in.

Above Carrara

for Paolo and Francesco

Climbing to Colonnata past ravines
    Squared by the quarryman, geometric gulfs
Stepping the steep, the wire and gear
    Men use to pare a mountain: climbing
With the eye the absences where green should be,
    The annihilating scree, the dirty snow
Of marble, at last we gained a level
    In the barren flat of a piazza, leaned
And drank from the fountain there a jet
    As cold as tunneled rock. The place –
Plane above plane and block on block –
    Invited us to climb once more
And, cooled now, so we did
    Deep between church- and house-wall,
Up by a shadowed stairway to emerge
    Where the village ended. As we looked back then
The whole place seemed a quarry for living in,
    And between the acts of quarrying and building
To set a frontier, a nominal petty thing,
    While, far below, water that cooled our thirst
Dyed to a meal now, a sawdust flow,
    Poured down to slake those blades
Slicing inching the luminous mass away
    Above Carrara . . .

Fireflies

The signal light of the firefly in the rose:
Silent explosions, low suffusions, fire
Of the flesh-tones where the phosphorous touches
On petal and on fold: that close world lies
Pulsing within its halo, glows or goes:
But the air above teems with the circulation
Of tiny stars on darkness, cosmos grows
Out of their circlings that never quite declare
The shapes they seem to pin-point, swarming there
Like stitches of light that fleck and thread a sea,
Yet unlike, too, in that the dark is spaces,
Its surfaces all surfaces seen through,
Discovered depths, filled by a flowering,
And though the rose lie lost now to the eye,
You could suppose the whole of darkness a forming rose.

Thunder in Tuscany

Down the façade lean statues listening:
Ship of the lightning-gust, ship of the night,
The long nave draws them into dark, they glisten
White in the rainflash, to shudder-out blackbright:
The threads of lightning net and resinew form
In sudden fragments – line of a mouth, a hem –
Taut with the intent a body shapes through them
Standing on sheerness outlistening the storm.

Giovanni Diodati

from the Italian of Attilio Bertolucci

My astonishment almost felicity
when I discovered Giovanni Diodati –
whose protestant Bible which I was reading
somehow entered my household – Catholic

if only tepidly with tenacious roots –
was the friend of that John Milton
whom today – late – I count among those poets
I care for most. The shimmer

of his lines – when he depicts Eve naked
garnishing a cloth
with reddening fruits in the autumn
of Paradise its noonday corruscating

at the guest's approach – Raphael
the Archangel – for a meal for three –
isn't it just the same as in the prose
of the exile from Lucca beside Lake Leman

where the Bride of the Canticle appears
suggesting to the intent adolescent –
fiery twilight coming slantwise in
to the resonant granary of wheat
hiding-place in air vertigo

of a plain black with swallows – the saliva of kisses?

On a May Night

after the prose of Leopardi's journal

Gloom in my mind: I leaned
at a window that showed the square:
two youths on the grass-grown
steps before the abandoned church
fooling and falling around
sat there beneath the lamp: appears
the first firefly of that year:
and one of them's up already
to set on it: I ask
within myself mercy for the poor thing
urging it *Go go* but he
battered and beat it low then turned
back to his friend: meantime
the coachman's daughter
comes up to a window
to wash a platter
and turning tells those within
Tonight it will rain
no matter what:
it's as black as a hat out there
and then the light at that
window vanishes: the firefly
in the interval has come round:
I wanted to – but the youth
found it was moving turned
swore and another
blow laid out the creature
and with his foot he made
a shining streak of it
across the dust until
he'd rubbed it out: arrived
a third youth from an alley-way
fronting the church
who was kicking the stones and
muttering: the killer laughingly
leaps at him bringing him down
then lifts him bodily:

as the game goes on
the din dies but the loud
laughs come volleying through:
I heard the soft voice
of a woman I neither knew nor saw:
Let's go Natalino: it's late:
For godsake he replies
it isn't daybreak yet: I heard
a child that must surely be
hers and carried by her
babblingly rehearse
in a milky voice
inarticulate laughing sounds
just now and then out of its own
quite separate universe: the fun
flares up again: *Is there any*
wine to spare at Girolamo's?
they ask of someone passing:
wine there was none:
the woman began laughing softly
trying out
proverbs that might fit
the situation: and yet that wine
was not for her and that
money would be
coin purloined from the family
by her husband:
and every so often she
repeated with a laughing patience
her hint *Let's go*
in vain: at last a cry
Oh look comes from them
it's raining: it was a light spring rain:
and all withdrew bound homewards:
you could hear the sound
of doors of bolts
and this scene
which pleased drawing me from myself
appeased me.

Instead of an Essay

for Donald Davie

Teacher and friend, what you restored to me
Was love of learning; and without that gift
A cynic's bargain could have shaped my life
To end where it began, in detestation
Of the place and man that had mistaught me.
You were the first to hear my poetry,
Written above a bay in Italy:
Lawrence and Shelley found a refuge once
On that same coast – exiles who had in common
Love for an island slow to learn of it
Or to return that love. And so had we
And do – you from the far shore of the sea
And I beside a stream in Gloucestershire
That feeds it. Meeting maybe once a year
We take the talk up where we left it last,
Forgetful of which fashions, tide on tide –
The buddha, shamanism, suicide –
Have come and passed.
Brother in a mystery you trace
To God, I to an awareness of delight
I cannot name, I send these lines to you
In token of the prose I did not write.

Barque Nornen

Barque Nornen broken by the storms
vanishes in shifting sand:
tides reshape the terrain and
unsilt the ship's clear hollow form:

Berrow church-tower looks out on
the ruin of that other nave
its sides like an inverted wave
and rib on rib as hard as stone.

The Littleton Whale

in memory of Charles Olson

What you wrote to know
was whether
the old ship canal
still paralleled the river
south
of Gloucester (England) . . .

What I never told
in my reply
was of the morning
on that same stretch
(it was a cold
January day in '85)
when Isobel Durnell
saw the whale . . .

She was up at dawn
to get her man off on time
to the brickyard and
humping up over the banks
beyond Bunny Row
a slate-grey hill showed
that the night before
had not been there . . .

They both ran outside
and down to the shore:
the wind was blowing
as it always blows
so hard that the tide
comes creeping up under it
often unheard . . .

The great grey-blue thing
had an eye
that watched wearily
their miniature motions as they

debated its fate
for the tide
was already feeling beneath it
floating it away . . .

It was Moses White
master mariner
owner of the sloop *Matilda*
who said the thing to do
was to get chains and a traction engine
– they got two from Olveston –
and drag it ashore:
the thing was a gift:
before long it would be
drifting off to another part of the coast
and lost to them
if they didn't move now . . .

And so the whale –
flukes, flesh, tail
trembling no longer
with a failing life –
was chained and hauled
installed above the tideline . . .

And the crowds came
to where it lay
upside down
displaying a
belly evenly-wrinkled
its eye lost to view
mouth skewed and opening into
an interior of tongue and giant sieves
that had once
filtered that diet of shrimp
its deep-sea sonar
had hunted out for it
by listening to submarine echoes
too slight
for electronic selection . . .

And Hector Knapp
wrote in his diary:
Thear was a Whal
cum ashore at Littleton Pill
and bid thear a fortnight
He was sixty eaight feet long
His mouth was twelve feet
The Queen claim it at last
and sould it for forty pound
Thear supposed to be
forty thousen pepeal to se it
from all parts of the cuntry . . .

The Methodist preacher
said that George Sindry
who was a very religious man
told himself when that whale came in
he'd heard so many arguments
about the tale of Jonah not being true
that he went to Littleton to
'satisfy people'. He was a tall man
a six footer
'but I got into that whale's mouth' he said
'and I stood in it
upright . . . '

The carcass
had overstayed its welcome
so they sent up a sizeable boat
to tow it to Bristol
and put it on show there
before they cut the thing down stinking
to be sold
and spread for manure . . .

You can still see the sign
to Whale Wharf as they renamed it
and Wintle's Brickworks became
the Whale Brick
Tile and Pottery Works . . .

Walking daily onto
the now-gone premises
through the 'pasture land
with valuable deposits of clay thereunder'
when the machine- and drying sheds
the five kilns, the stores and stables
stood permanent in that place
of their disappearance
Enoch Durnell still
relished his part in all that history begun
when Bella shook
and woke him with a tale that the tide
had washed up a whole house
with blue slates on it into Littleton Pill
and that house was a whale . . .

The Flood

It was the night of the flood first took away
   My trust in stone. Perfectly reconciled it lay
Together with water – and does so still –
   In the hill-top conduits that feed into
Cisterns of stone, cisterns echoing
   With a married murmur, as either finds
Its own true note in such a unison.
   It rained for thirty days. Down chimneys
And through doors, the house filled up
   With the roar of waters. The trees were bare,
With nothing to keep in the threat
   And music of that climbing, chiming din
Now rivers ran where the streams once were.
   Daily, we heard the distance lessening
Between house and water-course. But floods
   Occur only along the further plains and we
Had weathered the like of this before
   – The like, but not the equal, as we saw,
Watching it lap the enclosure wall,
   Then topping it, begin to pile across
And drop with a splash like clapping hands
   And spread. It took in the garden
Bed by bed, finding a level to its liking.
   The house-wall, fronting it, was blind
And therefore safe: it was the doors
   On the other side unnerved my mind
– They and the deepening night. I dragged
   Sacks, full of a mush of soil
Dug in the rain, and bagged each threshold.
   Spade in hand, why should I not make
Channels to guide the water back
   Into the river, before my barricade
Proved how weak it was? So I began
   Feeling my way into the moonless rain,
Hacking a direction. It was then as though
   A series of sluices had been freed to overflow
All the land beneath them: it was the dark I dug
   Not soil. The sludge melted away from one

And would not take the form of a trench.
    This work led nowhere, with no bed
To the flood, no end to its sources and resources
    To grow and to go wherever it would
Taking one with it. It was the sound
    Struck more terror than the groundlessness I trod,
The filth fleeing my spade – though that, too,
    Carried its image inward of the dissolution
Such sound orchestrates – a day
    Without reprieve, a swealing away
Past shape and self. I went inside.
    Our ark of stone seemed warm within
And welcoming, yet echoed like a cave
    To the risen river whose tide already
Pressed close against the further side
    Of the unwindowed wall. There was work to do
Here better than digging mud – snatching
    And carrying such objects as the flood
Might seep into, putting a stair
    Between the world of books and water.
The mind, once it has learned to fear
    Each midnight eventuality,
Can scarcely seize on what is already there:
    It was the feet first knew
The element weariness had wandered through
    Eyeless and unreasoning. Awakened eyes
Told that the soil-sacked door
    Still held, but saw then, without looking,
Water had tried stone and found it wanting:
    Wall fountained a hundred jets:
Floor lay awash, an invitation
    To water to follow it deriding door
On door until it occupied the entire house.
    We bailed through an open window, brushing
And bucketing with a mindless fervour
    As though four hands could somehow find
Strength to keep pace, then oversway
    The easy redundance of a mill-race. I say
That night diminished my trust in stone –
    As porous as a sponge, where once I'd seen

The image of a constancy, a ground for the play
    And fluency of light. That night diminished
Yet did not quite betray my trust.
    For the walls held. As we tried to sleep,
And sometimes did, we knew that the flood
    Rivered ten feet beneath us. And so we hung
Between a dream of fear and the very thing.
    Water-lights coursed the brain and sound
Turned it to the tympanum of an ear. When I rose
    The rain had ceased. Full morning
Floated and raced with water through the house,
    Dancing in whorls on every ceiling
As I advanced. Sheer foolishness
    It seemed to pause and praise the shimmer
And yet I did and called you down
    To share this vertigo of sunbeams everywhere,
As if no surface were safe from swaying
    And the very stone were as malleable as clay.
Primeval light undated the day
    Back into origin, washed past stain
And staleness, to a beginning glimmer
    That stilled one's beating ear to sound
Until the flood-water seemed to stream
    With no more burden than the gleam itself.
Light stilled the mind, then showed it what to do
    Where the work of an hour or two could
Hack a bank-side down, let through
    The stream and thus stem half the force
That carried its weight and water out of course.
    Strength spent, we returned. By night
The house was safe once more, but cold within.
    The voice of waters burrowed one's dream
Of ending in a wreck of walls:
    We were still here, with too much to begin
That work might make half-good.
    We waited upon the weather's mercies
And the December stars frosted above the flood.

Severnside

We looked for the tide, for the full river
 Riding up the expanse to the further cliff:
But its bed lay bare – sand
 That a brisk wind planed towards us.
Perpetual shore it seemed, stretch
 And invitation to all we could see and more:
Hard to think of it as the thoroughfare for shoals:
 At the edge, a cracked mosaic of mud,
Even shards of it dried in the sunny wind –
 A wind whose tidal sound mocked tidelessness,
Mocked, too, the grounded barges grass now occupied
 Dense on the silt-filled holds. Sad,
But a glance told you that land had won,
 That we would see no swell today
Impelled off the Atlantic, shelving
 And channelling riverwards in the hour we had.
And so we turned, and the wind possessed our ears,
 Mocked on, and our talk turned, too,
Mind running on future things,
 Null to all save the blind pull of muscle
In a relegated present. When we paused
 The sands were covered and the channels full:
We had attended the wind too long, robbed
 Of distinction between the thing it was
And what it imitated. But the rise we stood on,
 Reawakening our eyes, gave back suddenly
More than the good that we had forfeited:
 Ahead – below – we could sight now
The present, as it were, spread to futurity
 And up the river's bend and bed
The waters travelling, a prow of light
 Pushing the foam before them in its onrush
Over the waiting sand. And we who seemed
 To be surfing forward on that white
Knew that we only dreamed of standing still
 Here where a tide whose coming we had missed
Rode massed before us in the filled divide.

In the Estuary

This is the way it goes, the tide:
    A stain through the water, a first sign
That the light is getting down to layers
    Under the flowing surface: then
Colour brought up out of the depths
    To reveal suddenly a ledge of sand
Turning into a glassy island that reflects
    The further shore. The swimming birds
Are left standing, and walk to and fro
    Across their mirrored landscape, each
Accompanying its own white reflection.
    The channel shallowing, two rocks
(To begin with) and then a chain
    Of rock on rock, space out an archipelago
Of islets before the continental mass
    Which was the sand. You could not map
This making and unmaking. Every gap
    Is losing water. Every rise
Tussocked with grass that greens and fattens
    In the tidal flow, shows now
As an inland island above mudflats
    Through which the veins of channels hurry-off
What's left of the river from its bed.
    Reading the weather from such skies –
Cloud promise and cloud countermand –
    As cover this searun neither sea nor land,
You end in contraries like the bight itself,
    Where an unseen moon is pulling the place from focus
And the lunar ripple runs woven with the sunlight.

The Epilogue

It was a dream delivered the epilogue:
   I saw the world end: I saw
Myself and you, tenacious and exposed,
   Smallest insects on the largest leaf:
A high trail coasted a ravine
   Eyes could not penetrate because a wood
Hung down its slope: a fugue of water
   Startled the ear and air with distances
Around and under us, as if a flood
   Came pouring in from every quarter:
Our trail and height failed suddenly,
   Fell sheer away into a visibility
More terrible than what the trees might hide:
   Fed by a fall, wide, rising
Was it a sea? claimed all the plain
   And climbed towards us, smooth
And ungainsayable. We turned and knew now
   That no law steadied a sliding world,
For what we saw was an advancing wave
   Cresting along the height. An elate
Despair held us together silent there
   Waiting for that wall to fall and bury
Us and the love that taught us to forget
   To fear it. I woke then to this room
Where first I heard the sounds that dogged that dream,
   Caught back from epilogue to epilogue.